MEET THE WILD THINGS

Hello, I'm a LORIS

by **Hayley & John Rocco**

putnam

G. P. PUTNAM'S SONS

Oh, hello there. You startled me. I'm a loris, a pygmy loris to be exact. I get my name because I am the smallest of all lorises—about the size of your average squirrel!

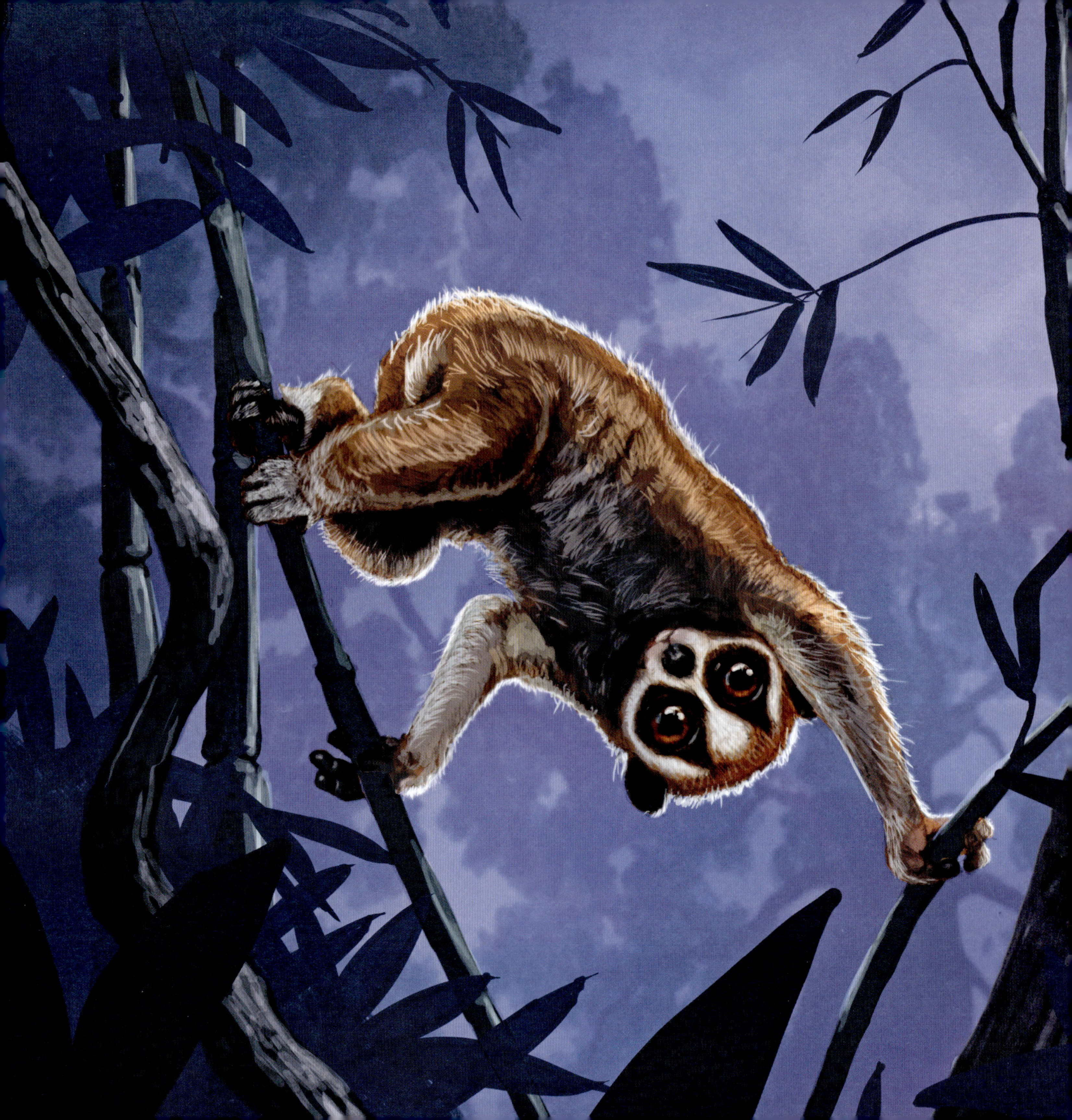

My strong arms and legs help me hang from these branches for hours on end.

Do you like hanging around?

I sleep during the day,
and I only come out when it is dark.

That's why I have such big eyes.
They help me see at night. In fact,
my eyes are so sensitive that bright
light can be almost blinding to me.

How well can you see in the dark?

I'm a primate, which means I'm related to monkeys, apes, and even humans. That includes you!

PRIMATES

gorilla

chimpanzee

tarsier

lemur

My favorite foods are tree gum and sap.
I nibble the gum that forms on the trunks,
and using my special teeth, I can make a hole
in the bark to drink the sap.

It's sticky and delicious, and the more of it I eat, the more a tree produces to replace it. This actually helps trees flush out deadly fungal infections and disease.

You're welcome, trees!

spotted lanternfly

Asian spongy moth

Asian longhorn beetle

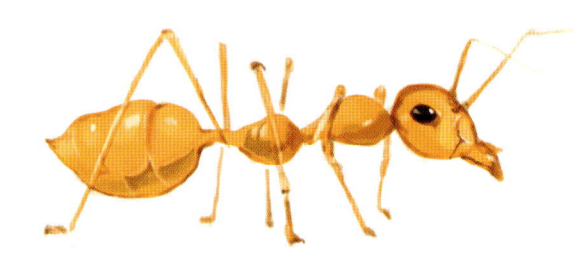

weaver ant

termite

I also help trees by eating a wide variety of insects that might hurt them.

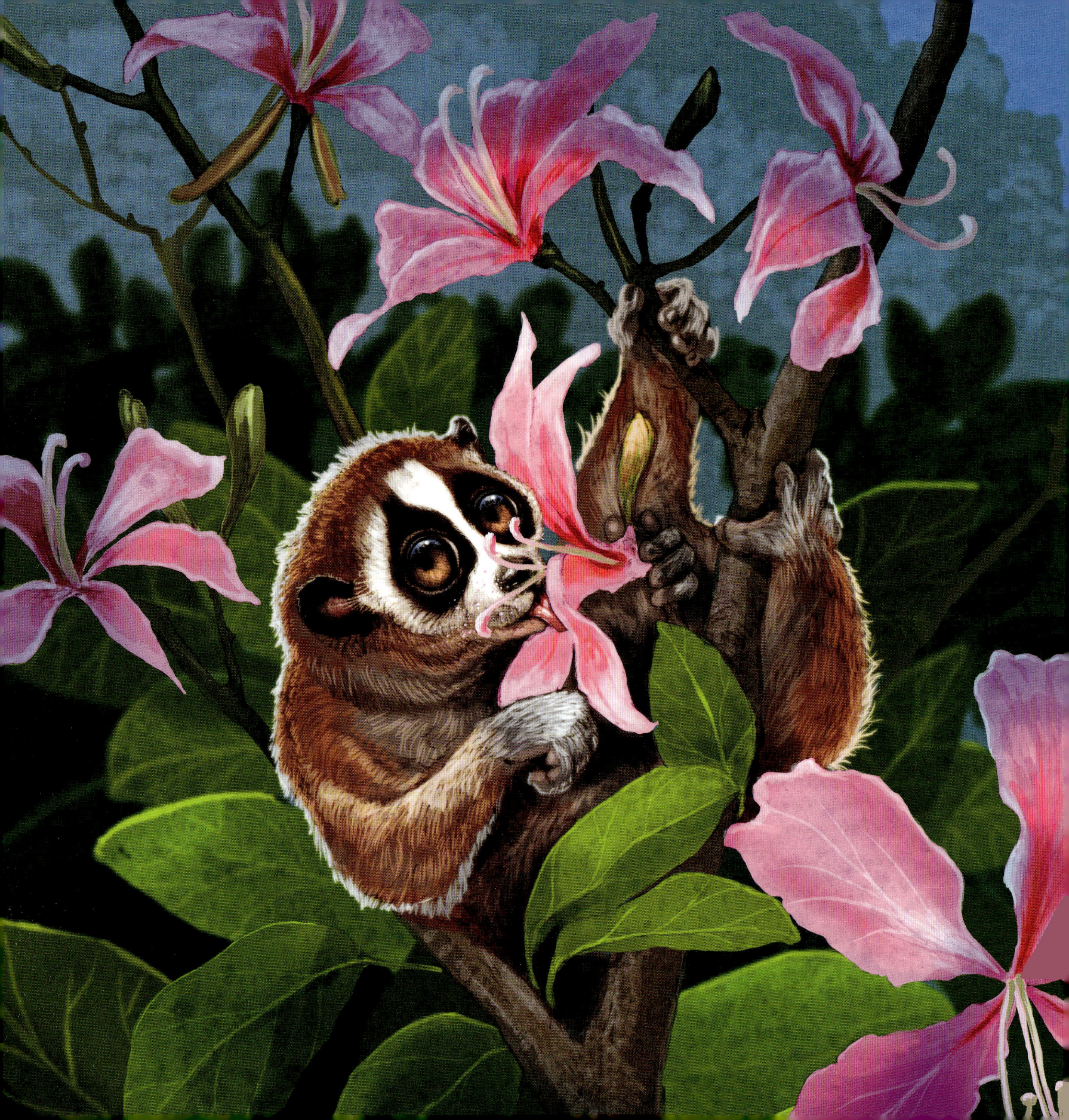

That isn't all I do to help my tall green friends. When I lick the sweet nectar from their flowers, I get pollen all over my face, which can pollinate the other flowers I visit. Just like a bee!

Do you eat anything that grows on trees?

ASIAN LORISES

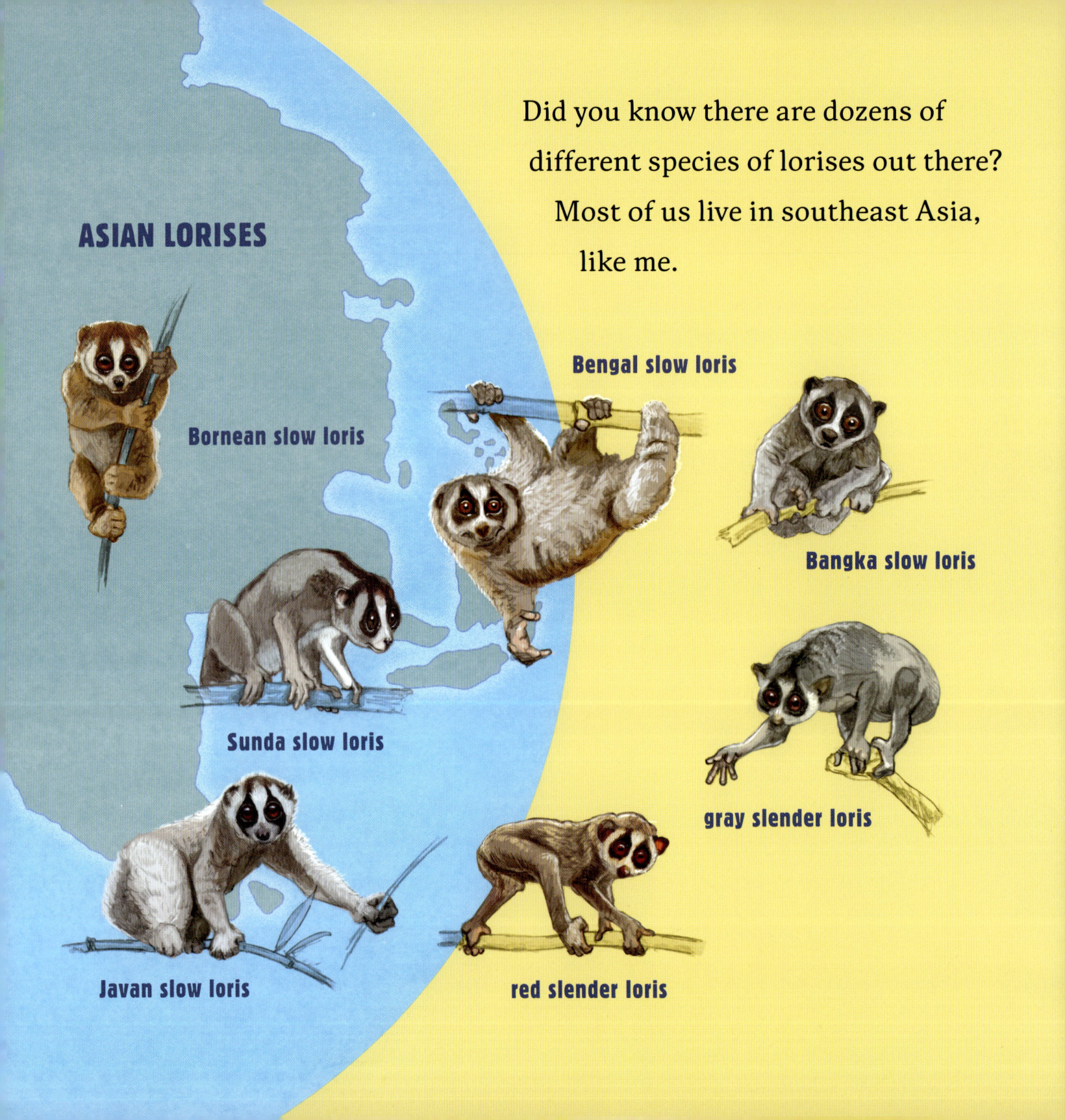

Did you know there are dozens of different species of lorises out there? Most of us live in southeast Asia, like me.

Bornean slow loris

Bengal slow loris

Bangka slow loris

Sunda slow loris

gray slender loris

Javan slow loris

red slender loris

Others can be found in some parts of Africa.

AFRICAN LORISES

West African potto

golden angwantibo

pygmy loris

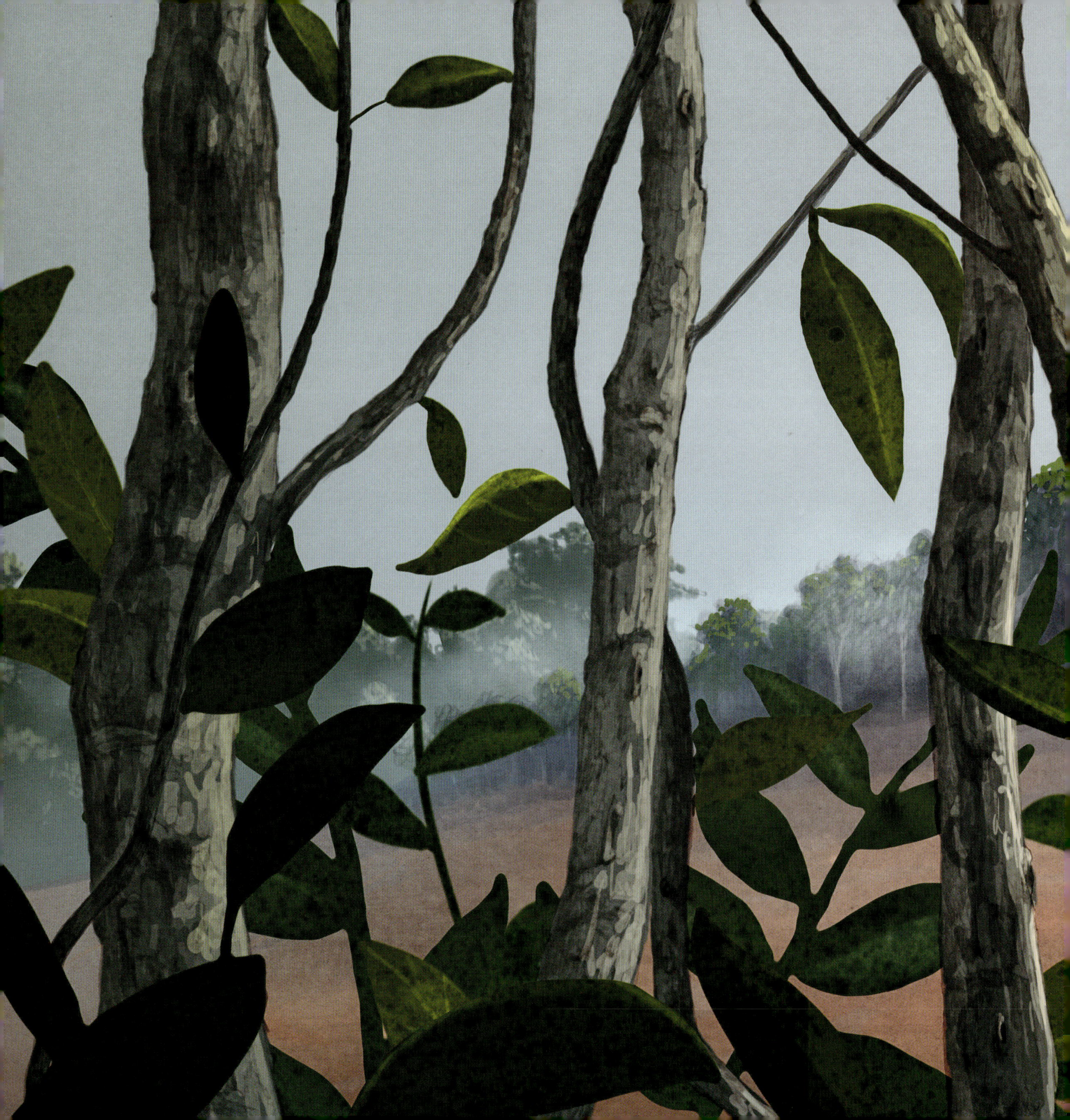

And believe it or not, we are all endangered.
The main reason is our rainforest habitats are being
destroyed. The trees are cut down for logging,
or the land is cleared for farming and development.

Humans also take us from our homes to sell as pets.
Kittens and puppies and even goldfish make great pets.
I, on the other hand, do NOT make a good pet.

I may look cute and cuddly, but that doesn't mean I want to be snuggled. Unlike puppies, I don't like hugs, belly rubs, or tickles.

Do you like tickles?

When I am afraid, I growl and hiss.
But if that doesn't work,
I may have to bite.

I'll tell you a secret: I am the only primate in the whole world that is venomous! If I feel scared or threatened, I release a toxic oil from special glands on my upper arms. When I lick it up, it mixes with my saliva to make a potent venom. Then I can deliver a nasty, painful, and sometimes even deadly bite.

But don't worry, I really am a nice loris.
Sometimes I just need my own space.

Do you need
your own space
sometimes, too?

There are some humans who are trying to protect lorises like me. They rescue us from the illegal pet trade, and when we're ready, they release us back into the wild. They also teach other humans about us.

Will you tell others about us, too?

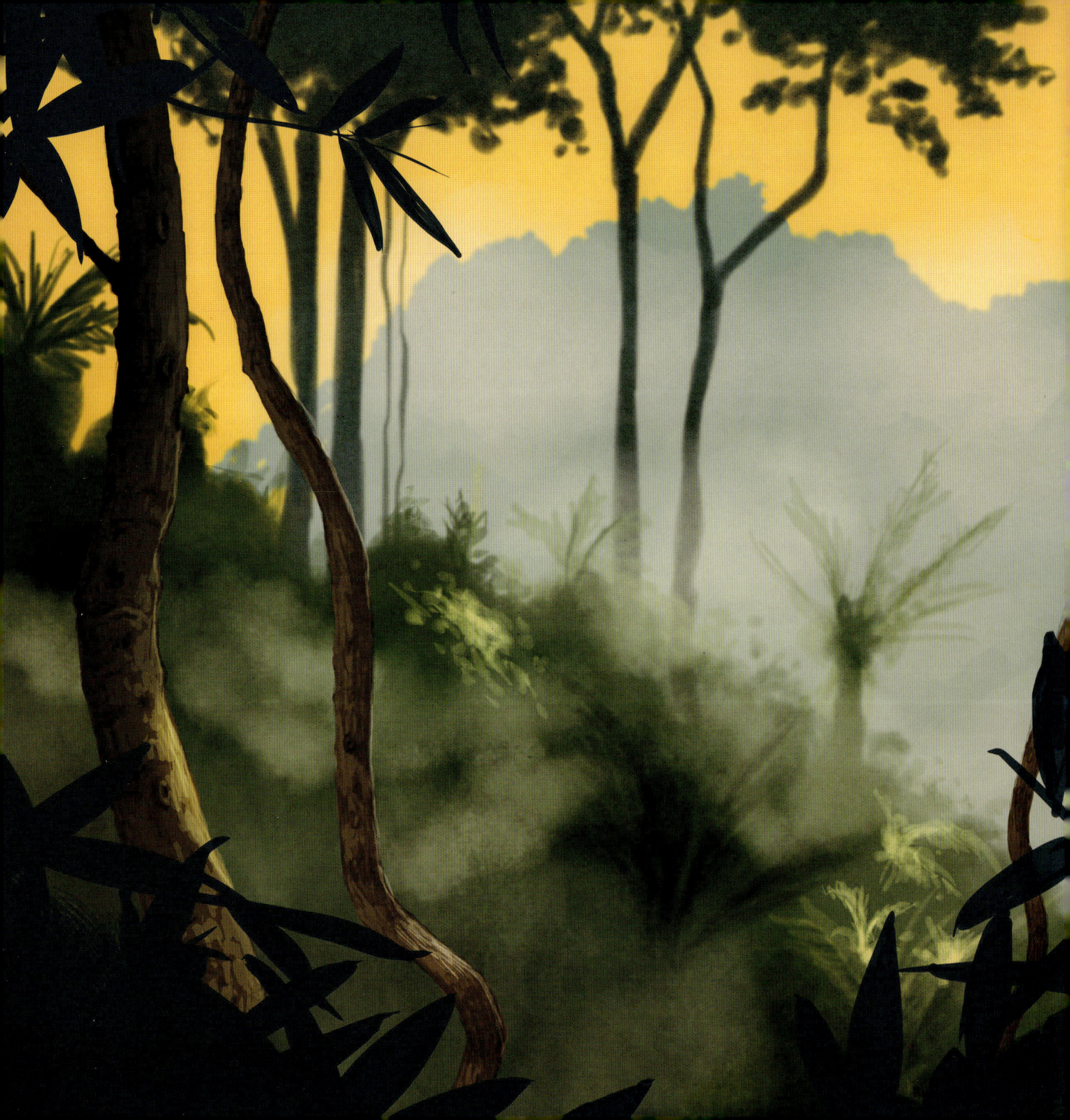

Well, it's time for me to find a comfy place to curl up and sleep. It'll be daytime before you know it.

Goodbye!

A little more about lorises:

- The pygmy loris is the smallest of the loris family at about 6 to 10 inches long. The largest is the Bengal slow loris at about 10 to 15 inches long.
- Two species of pygmy lorises (northern and southern) live in forests east of the Mekong River in Vietnam, Eastern Cambodia, and Laos.
- The pygmy loris lives between 10 and 20 years. The oldest known pygmy loris lived for 22 years.
- It was previously thought that lorises were solitary in the wild, but in recent research, pygmy lorises have been observed traveling in small family groups.
- Pygmy lorises use their venomous bite to settle disputes over territory with other lorises. They also groom themselves using their venom to ward off predators and parasites.
- Pygmy lorises spend most of their lives in trees and rarely go to the ground. They can hang from branches with their strong feet while they gather food with their hands.
- About 50–70 percent of the pygmy loris's diet is tree gum and sap.
- Lorises are nocturnal, meaning they are most active at nighttime and usually sleep during the day.
- In summer months, they stay active, constantly moving throughout the night. But in winter months, when there is less food available, they spend most of the day (up to 19 hours!) sleeping. They also can slip into a shortened type of hibernation called "torpor," which helps them conserve energy.
- About six months after mating, a mother gives birth to one or two babies at a time. It is common to have twins!
- When loris babies are first born, they cling to their mothers' bellies as they forage for food and travel throughout the rainforest. Once the babies are a bit older, their mothers leave them in a safe place for short periods of time while looking for food.
- Mothers nurse their babies for about 18 weeks.
- Young pygmy lorises stay with their mothers for up to three years before they have learned enough to live on their own.
- Lorises are one of the most understudied primates on earth, so scientists are still learning about how they live in the wild.

Why are lorises endangered?

Like most endangered species, habitat loss due to human activity is one of the biggest reasons lorises are endangered. In the region where pygmy lorises live, the rainforests were also devastated by fires and deforestation during the Vietnam War (1955—1975) and other human conflicts. Additionally, lorises are poached—or illegally hunted—for their meat, for use in unproven traditional medicines, and for the illegal pet trade.

No wild animals should ever be taken from their habitats to be kept as pets, and lorises are no exception. Once they are captive, lorises do not have access to their specialized diet of tree gum and sap, and their health deteriorates. And in order to sell them as pets, poachers remove lorises' sharp teeth in a very painful procedure.

Special wildlife authorities investigate the illegal pet trade and work to catch wildlife poachers. Unfortunately, the popularity of the loris on social media has only increased demand for them in the illegal pet trade. By choosing not to interact with any captive loris content on social media, you can help keep lorises in the wild, where they belong.

Organizations working to help lorises:

Little Fireface Project: Nocturama.org
International Animal Rescue: InternationalAnimalRescue.org/projects/slow-loris
Endangered Primate Rescue Center: EPRC.Asia/rescue-rehabilitation/project-loris-rewilding
WildAct: WildAct-production.vercel.app

pygmy loris

©K.A.I. Nekaris

For more information about lorises and how you can help them, visit
MeetTheWildThings.com

For my uncle Noldo, who loves the pygmy loris and took me
to see a very special family of them at the National Zoo. —H.R.

For Scott Douglas. —J.R.

HAYLEY AND JOHN ROCCO are both ambassadors for Wild Tomorrow, a nonprofit focused on conservation and rewilding South Africa. They are the author and illustrator team behind the picture book *Wild Places: The Life of Naturalist David Attenborough*. John is also the #1 *New York Times* bestselling illustrator of many acclaimed books for children, some of which he also wrote, including *Blackout*, the recipient of a Caldecott Honor, and *How We Got to the Moon*, which received a Sibert Honor and was longlisted for the National Book Award. Learn more at MeetTheWildThings.com.

ACKNOWLEDGMENTS Our immense gratitude goes to Prof. Dr. Anna Nekaris, the world's leading expert on Asian lorises, director of the Little Fireface Project, and vice chair of the Special Section of the IUCN Primates Specialist Group for African and Asian Prosimians. Thank you for your expertise and guidance to ensure our information was as accurate as possible for the making of this book. We also want to thank Katey Hedger, Indonesian project leader and research coordinator for the Little Fireface Project, for connecting us with Prof. Dr. Anna Nekaris and for her dedication and hard work protecting these animals in the wild. Finally, thank you to all the conservationists working so hard to not only protect these creatures, but also educate the world about why they're important to our planet.

G. P. PUTNAM'S SONS | An imprint of Penguin Random House LLC | 1745 Broadway, New York, New York 10019
First published in the United States of America by G. P. Putnam's Sons, an imprint of Penguin Random House LLC, 2025

Text copyright © 2025 by Hayley Rocco | Illustrations copyright © 2025 by John Rocco

Penguin Random House values and supports copyright. Copyright fuels creativity, encourages diverse voices, promotes free speech, and creates a vibrant culture. Thank you for buying an authorized edition of this book and for complying with copyright laws by not reproducing, scanning, or distributing any part of it in any form without permission. You are supporting writers and allowing Penguin Random House to continue to publish books for every reader. Please note that no part of this book may be used or reproduced in any manner for the purpose of training artificial intelligence technologies or systems. | G. P. Putnam's Sons is a registered trademark of Penguin Random House LLC. | The Penguin colophon is a registered trademark of Penguin Books Limited. | Visit us online at PenguinRandomHouse.com.

Library of Congress Cataloging-in-Publication Data is available.
ISBN 9780593858783 | 10 9 8 7 6 5 4 3 2 1
Manufactured in China | TOPL

Design by Nicole Rheingans | Text set in Narevik | The art was created with pencil, watercolor, and digital color.

The publisher does not have any control over and does not assume any responsibility for author or third-party websites or their content.

The authorized representative in the EU for product safety and compliance is Penguin Random House Ireland, Morrison Chambers, 32 Nassau Street, Dublin D02 YH68, Ireland, https://eu-contact.penguin.ie.